The Fate of Humanity

Do Cosmos and Chance Really Affect our Fate?

By

Dr Ugur GUVEN

Guven Publications

Miami-Paris-Istanbul

Disclaimer

This is a work of fiction. Names, characters, businesses, places, events, locales, and incidents are either the products of the author's imagination or used in a fictitious manner. Any resemblance to actual persons, living or dead, or actual events is purely coincidental.

Although the author and publisher have made every effort to ensure that the information in this book was correct at press time, the author and publisher do not assume and hereby disclaim any liability to any party for any loss, damage, or disruption caused by errors or omissions, whether such errors or omissions result from negligence, accident, or any other cause.

Copyright

The Fate of Humanity
© January 2020, by Dr. Ugur GUVEN
Self-published through Amazon under Guven Publications

ASIN: B083LCV2D1
ISBN: 978-1657181144
Miami-Paris-Istanbul
Contact: drguven@live.com

All rights reserved.
No part of this publication may be reproduced, stored in a retrieval system, stored in a database and / or published in any form or by any means, electronic, mechanical, photocopying, recording or otherwise, without the prior written permission of the publisher.

Foreword

This is my first attempt at writing a non-fiction book about where the future of humanity is going forward to. Though this particular work is around 10,000 words so it is more like a handbook that examines the particular forces in human history that effects its fate and then I have tried to project what the future of humanity will be based upon these particular forces. As a Rocket Scientist with a PhD, I have worked on many research papers and textbooks and they are sold across the world. I have especially worked on interstellar travel techniques and if you go to my website, you can see and read many of the research papers that I have worked on. This book specifically focuses on the factors that effected the past development of humanity such as entropy and time and chance. I have also tried to paint a picture of what the future would look like.

While many of us don't realize it, but actually humanity is slave to various vagaries of time and as a subset of civilization, we as individuals are also slave to these same constraints. However, as bleak as things may look, we do have a better future then we think, and I also think that as individuals we can reshape our destiny to some extent. I am from the 1970's generation and I am pretty sure that I speak for everyone from my age group, that I think we were born a century too early.

Please do share your feelings with me after reading this book and I hope that my science fiction writing will improve over time.

I dedicate this book to the future of humanity.

Prof. Dr. Ugur GUVEN

www.drguven.com

SUMMARY

This book examines the various forces that exist since the beginning of time which has affected the history and the development of human civilization from one way or the other. These forces include various forces such as entropy, gravity, chance and many others that you will discover while reading my book. These forces have swayed our history by limiting various factors and forcing a particular behavior type for humanity. Based upon the understanding of these forces, it's possible to predict the future of humanity and to see how we can best prepare as humanity to the glorious future that may be awaiting us. In addition, understanding these forces will help you to reign in these forces for your individual destiny as a person.

The Fate of Humanity

Introduction: L'Reve Des Toiles – The Dream of Stars

Mankind has been looking toward the stars since the dawn of civilization. L'Reve Des Toiles or the Dream of Stars has been one of the most fundamental forces driving mankind since millenniums and it is perhaps responsible for many of the discoveries that we face today. Even as we look back at the ancient Greeks and ancient Mayans, we will see the effect of stars on the development of their civilization as it lies at the root of their accomplishments. Ancient Greeks invented Physics based upon the need to understand the cosmos and furthermore the Ancient Mayans were masters of Astronomy as they believed that the working mechanism of the Cosmos has a direct bearing on our daily lives. If we go back as far as Ancient Sumerians, we will see that they actually had the workings of a basic string theory of physics even back then. Since the first sign of civilization in Göbeklitepe in the ancient Mesopotamian region, cosmos and the heavens has pushed mankind to learn more about the universe and to find ways to control the elements of the universe as much as they can. Göbeklitepe is considered as the first civilized town that has been unearthed as the first signs of devotion to Gods was seen in Göbeklitepe as well as organized homes and an organized village.

Fig 1: Göbeklitepe – The Cradle of Civilization

Even today while thousands of years have passed since the inception of Göbeklitepe, humankind and civilization have the same traits in many ways. If we look at the civilization in general, we will see that horoscope and religion are also another reflection of this concept, as people try to find relations between the cosmos and their fate as well as their future. This basic driving force has been definitive of many actions of billions of human beings over the centuries.

What is it that pushes mankind to learn more about the cosmos and to try to understand his fate?

Is it the dream of stars that has such an effect on mankind, as perhaps it can be considered as a romantic overture which helps to understand the place of humanity in the greater scheme of things? Or is it perhaps the fact that humanity is interested in controlling his fate and he inadvertently believes that can only be done by controlling the elements of the universe. If we look at the psychological aspect of it, we can see that the belief in Astrology also is based on this premise that by understanding the dynamics of the Cosmos, we can predict and perhaps influence our future. Thus, it can be said that while Astrology may not be accurate, there does seem to be some basis as to why cosmos can have an effect on the future of mankind as a whole if not the individuals.

So, then how do we start the journey to explore and to understand the fate of humanity as a whole and even fate as a single person? Isaac Newton, which is perhaps one of the 3 great scientists of all times, is said to have found the equation of the Universe and that with that particular equation it was possible to predict everything in the universe including the future. His premise was that if you know the position of every particle in time, then you will have an answer to everything that happens in the universe including knowing the future of everything and everyone. The legend has it that he was so happy with this finding, he went to a nearby church to pray his thanks to God, but by the time he got back, his laboratory was in flames and his equation was lost to time forever.

It seems that there are certain forces at work that directly affect the rules of the universe and these must be understood before we can decide on the fate of the Humanity. At the end of the book, you will perhaps see that the future of humanity may also determine your future as an individual and understanding the cosmos may be more than just an intellectual exercise as it may help you to unlock your future as well as your fate.

CHAPTER I: THE ARROW OF TIME

When Stephen Hawking published his popular book '*The Arrow of Time*", it created a lot of interest in general public. The reason for this stemmed from the fact that Stephen Hawking for the first time showed how interesting science can be and how important some scientific discoveries are for the understating of the universe. He showed that a complex concept such as Blackhole Physics can be understood by the general public and that it can be a way to open up our understanding of our universe as well as our future.

Hence, we understood that the arrow of time or the vector of time is not just any variable in physics, but perhaps it is one of the most important concepts in understanding our universe and hence our fate. This is perhaps very important to understand because if we look at it, we will see that the presence of time has a direct effect on our lives, and it is not just an abstract physical constant to be used in equations.

Fig 2: Arrow of Time always Pointing Toward the Future

First of all, from both physics and daily life point of view, time is something that only seems to flow forward as you can never get back to it downstream. It is like a river that flows and once you pass a certain point, you can never get back to that point again. This is perhaps to be expected as any moment that you live, as even after a second it becomes just a memory and while you can revisit your memories, you can never revisit a particular point past in time. Of course, popular movies such as *"Back to the Future"* romanticizes the concept of going back in time or popular novel *"The Time Machine"* by HG Wells tells us that time travel can be dangerous as well. From a physics point of view, time vector does seem to point forward in time. Once you live a particular moment, no technology exists for you to revisit that moment or to live that moment again perhaps with different decisions. Thus, this creates an interesting conundrum as it creates a natural barrier for the mankind, as all activities must be planned in such a way that they can never be revisited or redone again. Maybe many people may not realize it, but this itself is something that has shaped the civilization of mankind, since knowing that you can only move forward in time changes the way you think and the way you do things.

Imagine that such a world exists where you can go back and change your decisions and actions if you feel you need to redo it in order to get a better outcome. You can quickly see that everything from wars to revolutions would change very quickly and unless it is controlled, you can have a chaos in a very short amount of time. Of course, there would be no possibility of failure, as you could go back and change that failure into success at any time. Wars would become obsolete as you could always go back and change your mistakes to help win any war, but obviously this would work both ways both parties could do the same so winning a war would become effectively impossible. Hence, eventually the parties would realize that fighting is pointless when there is no concrete ending to the war. Naturally, deaths due to accidents and even due to medical problems can prevented, as you can go back in time and change the parameters, so that the natural death does not occur for that person.

However, imagine the consequences of such a world. The resources of the world would be depleted very fast and the people may not learn from their mistakes as they will have the luxury of going back again and again to change those mistakes which may create lethargy on the part of the society. For example, imagine a student who gets an F because he does not study for the exam. If the student can go back and retake the exam again and again, there would be no incentive to study. Furthermore, it would be impossible to have the sanctity of the exam as anyone can go back in time and see the questions and know the answers before taking the exam. Or as an example imagine a burglar who is caught while trying to rob something. He or she can go back in time again and again until he is successful. Hence, while it may sound very nice to have time travel capability, if it was a given fact that the time vector can be used to move backward in time, it would probably have disastrous consequences and cause extreme chaos and end to society for the reasons detailed above.

So, it's possible to imagine how the arrow of time actually has shaped mankind in a certain way to compete for resources and for success in his daily endeavors. Thus, it won't be a mistake to say that the arrow of time has an infinite effect on the fate of the humanity.
Of course, there have been many science fiction books and movies which has focused on beings that don't see time as a singular arrow in one direction and can manipulate time. However, as far as today's technology

is concerned, these beings would be magical and possibly God Like as far as we are concerned.

CHAPTER II. THE DRAG OF GRAVITY

If the Arrow of Time seems to be the most important factor in the fate of humanity, then perhaps Gravity would be the second most important factor. Anyone who has watched the movie Interstellar can appreciate how much of a role the concept of gravity can play in humankind's future. Since the dawn of time, gravity has also been an important influence, as it has caused mankind to stay firmly rooted on Earth and that has also had an important bearing on the way humanity has behaved.

For example, all of the wars that have been fought in history including the two world wars have been fought only because of the competition for resources, since Earth has finite amount of resources including land, oil and even water. Thus, knowing that you are limited to a finite amount of resource is a driving force for wars and political conflicts all across the world. Hence, the concept of gravity has determined the fate of nations and people all around the world in an indirect manner.

Fig 3: Gravity – Unseen Force All Around Us

If we look at the present situation, we will see that it is becoming worse, since more and more resources are dwindling and with the growing population of Earth, it is becoming increasingly difficult to handle these growing needs including food, water, energy, oil and mineral resources. There is always the issue of a potential outbreak of some resistant virus (like the outbreak of the Spanish Flu in the early 20th century) and several other issues due to global warming, which has been actually caused by the competition for these resources.

In fact, experts agree that global warming is seriously becoming such an issue that many places on Earth may become uninhabitable in the next few decades. The places that used to get much rain are facing draught and the places that used to have drought are having floods. There are tornadoes and hurricanes in places that never had them before and especially since 2018, the hurricanes are coming with more devastating force than ever. The places which has a desert climate is facing torrential rainfalls and the regions which typically received very less precipitation are facing severe drought. In regions where there is winter, the snowstorms are taking place in intensities never seen before causing even airport and other transportation options to become unavailable to the public. Thus, global warming in addition to dwindling resources means that mankind's future fate is in peril and needs to find alternatives, so that survival as a species as well as a survival at an individual level becomes possible.

As long as gravity exists, it would be impossible to have a situation where leaving earth would be easy and cheap. In fact, not only for leaving Earth, but also for easy transportation it is a challenge. The fastest way to travel is by plane and flying a commercial jumbo jet is not a small feat and it is quite expensive as an aircraft needs to have a crew and it also needs to be maintained 24&7 for it to be commercially viable. If some day mankind conquers gravity, then Earth will become the cradle of humanity as people can transport themselves around the Earth and into the solar system with very little effort or cost. It can easily be seen that this would be a grand game changer for the humanity as a whole and it would affect the quality of life of each individual being as well. While it may be just a distant dream for many people, we may be closer than ever to such a breakthrough.

CHAPTER III: CHANCE - DOES DICE HAVE MEMORY?

Most casino owners would like to boast that dice has memory in casinos and as a result, you should keep playing until you can possibly win. Is this true? Does dice have memory? Eventually it can be said that statistics and random events have an important effect on the date of humanity. Statistics is without a doubt one of the most important branches of mathematics that has an effect in all branches of science. In fact, without statistics, it will be very difficult to explain many concepts in the universe which actually depend on the ordering of random events. Eventually, statistics deals with ordering and understanding of random events as well as non-random events which may have causality as an effect. For example, if you flip a coin, you have 50-50 chance of a coin coming up either as heads or tails. It is a random event and no matter how many times you flip a coin; it will always come up as either heads or tails with 50-50 chance at every flip that you do. However, there is also the law of averages that states that each randomized event will average out if you have a big enough sample. Hence, for example if you have only 20 flips, overall you may not get 50% heads and 50% tails, but if you do it 10,000 times or 100,000 times then that chance for getting a perfect 50-50 toss grows. So, is it possible to say that fate of humanity can be predicted based upon this? To understand this, we need to examine something called a Gambler's Fallacy.

Fig 4: Chance – Do Dice Have Memory?

Basically, Gambler's Fallacy lies on the assumption that each randomized event actually has a memory and that if a certain event has happened too many times in a row, there is a good chance that it won't happen again and thus you should choose the alternative outcome. The concept comes from an actual gambling scenario, where if the crap table has had red coming 10 times then it should possibly not come red again the 11th time. As another example, if the dice has come as 6 couple of consecutive times, then the dice won't (or should not) come as 6 again. This depends on the notion that dice have memory and that the chances of 6 coming again is very low, as the universe will balance it out according to the law of averages. From a statistical point of view, this is incorrect as the chance of a random event happening at each cycle is always the same no matter how many times that event has been repeated consecutively in the immediate past. That means that when the coin has come as tails 10 times in a row, the chance of a coin coming a tail on the 11th run is still 50-50 and it will not be less due to some suggested memory of the coin or due to the law of averages. This can be applied in all aspects related to random events.

If we look at the major world wars, we can see the effects of Gambler's Fallacy in world history. For example, the two major wars started in Balkans and technically as per Gambler's Fallacy, many experts at the time didn't believe that Balkans could be causing such activity, but eventually it did.

If we look at all the wars in the last 150 years, we will see that they are all energy based wars as various nations fought each other to control resources. Thus, it seems there is no Bayesian connection to past events, as each war seems to be fought for the same reason and chance seems to favor this as well.

So, we come back to the same premise to understand the factors that control the fate of humanity. Does dice have memory? Can we say that if a random event has happened many times that it won't happen again? Strangely the answer to that question is yes and no at the same time.

Take an example of a family that has history of cancers in their past. Imagine that a member of the family has cancer and that it has a devastating effect on the overall happiness of the family. However, the family members may think that the worst is over as statistically the chance of getting cancer has already been met once, only to find out that another member of the family has cancer. Of course, sickness also leads to financial miseries as well as spiritual ones and the world is full of cases where a medical problem caused the family to fall apart and every member of the family kept suffering due to cascade of events that kept continuing. In a way, it can be said that regardless of chance, misery gives birth to misery and as such it defies the laws of statistics.

In many parts of the world, there is a saying that money and success attracts money and success. If you have money and success, many people will attest to the fact that it will attract more money and more success and many people with financial difficulties will probably face more and more difficulties as time passes.

When you see a very poor family then also having a terminal illness may seem a small chance due to Gambler's Fallacy, but in reality, and in all likelihood a poor family will also face serious health problems including terminal ones. Of course, it is to be expected as a poor family will have a less healthy lifestyle which may cause a cascading effect. Hence, people with misery will probably face more misery, while persons with money and means may have better and better lives.

Naturally, there may be situations that don't fit this, but law of averages will usually come out in favor of this premise. In a way, it can be said that dice do have a memory and that memory favors the actions of the past, but at the same time there is always a 50-50 chance that the outcome may be different this time. However, many gamblers will proclaim that when they are on a winning streak, they will keep winning until they are on a losing streak again. Many people will attest to the fact that when you are on the losing streak, you will continue to keep losing until eventually it comes to an end one way or the other.

While some people believe in the hypothetical law of attraction, which basically says that positive things and thoughts positive energy and thoughts, while negative things attract negative thoughts and energy. Especially if you have watched the movie "Secret" then this may be a strong belief. While this may have some truth to it as we will examine in a separate chapter, actually the laws of statistics do prevail over humanity.

If the resources are low, then the probability of conflicts arising from those low resources will always be there. Furthermore, law of averages will demand catastrophic events both natural and manmade to take place for controlling the availability of resources. However, there will always be a chance that the event may have a different outcome (50-50 chance) at every event. Though it may sound contradictory to say that dice have a memory and a 50-50 percent chance of having an outcome, actually both concepts fit well with each other and it is one of the events in the universe that creates hope in spite of any bad experiences in the past. It actually gives a driving force to the humanity that regardless of how bad things may have been in the past or how much the chance may favor a negative outcome, there is always the chance that the dice may come up differently this time. In fact, in a way this has been a major differentiator for change for centuries in mankind and will continue to be so.

CHAPTER IV: ENTROPY – DOES CHAOS CONTROL MANKIND?

One of the most interesting concepts in physics and especially in thermodynamics is the concept of entropy. Basically, it can be described as the amount of chaos in the system and one of the tenets of entropy is that the universe always goes towards a state of disorder or maximum entropy. For example, imagine that you have 10 red balls and 10 blue balls and assume that you mix them together in a container. You will see that you will have random collection of blue balls and red balls and they will never be ordered on top of each other by themselves. Same thing, you can think of any ordered system in the universe. Due to entropy all ordered systems will eventually disintegrate into a system of chaos and thus it's not possible to return to the original state in a random manner. From chemistry to physics, entropy shows itself as friction, dissipation of energy, loss of efficiency etc. and it is the reason why there is no perpetual motion machine. Otherwise you could have made a machine that starts once and never stops working until eternity even when you don't add energy or work into the system.

Fig 5: Entropy – Can Chaos be Ordered?

So, then you may think how this concept applies to the whole of the mankind. Actually, entropy plays a more important part then many people give credit for. Take life for example, as life will depend on series of events to take place, however once you are born, then you are subjected to various processes that will try to upset the balance of chemicals and biological processes in your body. Eventually, even if you keep yourself fit and as healthy as possible, eventually your chemical and biological processes in your body will decay and eventually you will die due to the loss of efficiency in the biological and chemical processes. Thus, we see the hand of entropy in everything that involves life and in a way, it is the reason why all living being die and thus have a finite life span.

Many philosophers and historians agree that this concept of having a finite life span due to entropy has a profound effect on the fate of humanity as a whole and on the fate of an individual as a lone unit of existence. Imagine that you are in a position of power and you know that you have a finite life span. Many world leaders have seen this conundrum and have taken steps to do activities that will show their greatness in their finite lifetime. This has both positive and negative effects on the situation, as some leaders will use this limited time frame to create wars to conquer lands and resources to show their greatness like Alexander the Great once had done. Furthermore, if you look at many influential leaders like Caesar, Constantine, Napoleon and others, they have always acted with the instinct that was detailed above. On one hand, the concept of having a finite life span has helped many cultures to reach greatness through the individual efforts of leaders that wanted to establish themselves in a finite amount of time. In fact, if we look at the way great world empires and civilizations were formed, we will see similar symbolism in the Greek Empire, Roman Empire, Ottoman Empire and the British Empire where at one point there was no sunset throughout the empire due to its vast size.

Of course, the same concept can also work against the civilizations as sometimes too many wars and conflicts can lead to famine and loss of political stability. In addition, having a finite life span also stops leaders from being able to finish all the necessary reforms, because these reforms usually can take more than a lifetime. In these instances, the concept of having a finite lifetime may have an exact opposite effect against progress and thus it may even lead to stagnation of civilizations as it can be seen in the cases of the Roman Empire and Ottoman Empire.

Naturally, the concept of entropy has other repercussions on the existence of humanity then just introducing the concept of a finite life span. In fact, the existence of entropy as mentioned before is one of the reasons why we can't have a perpetual motion machine, as it the presence of internal dissipation due to friction would always stop a system from running forever. For example, imagine that you are riding a bicycle. No matter how hard you press on the pedals, eventually the bike will come to a stop if you stop pedaling the bicycle. This will be due to the combination of both the friction caused by the internal gears of the system as well as the friction the tires will have with the road itself. You can see how this will have an effect on any physical system created by mankind as this will cause mankind to create systems that constantly require energy to run which creates the energy deficiency that has become an integral part of the global economy.

Imagine that cars could have run forever once started or planes that could fly forever with just the initial startup momentum or imagine factories where the machines run only by requiring initial startup power. You can easily imagine that the world economy will become very different as the world resources can be instantly freed. This is a magnificent event, as the need for energy will die down suddenly and it will also stop many conflicts from happening including wars across the world. No one would need to compete with anyone else for energy resources or for any resources that require energy to work. This would be very utopic world and theoretically you can have a world that doesn't have much wants or needs that can't be satisfied as they will have access to many things with very little effort or cost. Of course, this would also cause an unchecked population expansion across the world, which may have unintended consequences for the world as a whole. Unfortunately, everything in the world corresponds to the limitations of entropy and as a result, every system requires a constant input of energy, which causes constant need to search for energy sources. This search for energy will be examined in detail in the next chapter, but eventually the need for energy stems from entropy and it can be said that entropy is a very important factor in limiting humanity's fate and yet at the same time driving it forward to overcome these limitations. Hence, in order to determine the fate of humanity, entropy needs to be factored in as it effects our life spans as well as our ways and means since everything gets affected by entropy in the economic system. In a way, there are some economic theories that suggest that the existence of entropy and its effects indirectly effect the cost-benefit balance in the world and effect the economy of the world that we live in. The amount of money and effort that spend to overcome entropy also determines the overall result of our fate as a person and also as species.

CHAPTER V: ENERGY – ENERGY CAN NEITHER BE CREATED NOR DESTROYED

One of the most important rules of physics that govern the universe is that energy can neither be created or destroyed, it can only be transformed into another form. This fundamental tenet has shaped our understanding of the universe. The universe has a total amount of finite energy, and for all physical processes, this finite energy needs to be used. Everything in the universe from the creation of stars to functioning of solar systems depend on this energy. On the planetary level, every physical reaction on the planet ranging from rain to the growing of crops depends on the dispersal of this finite amount of energy.

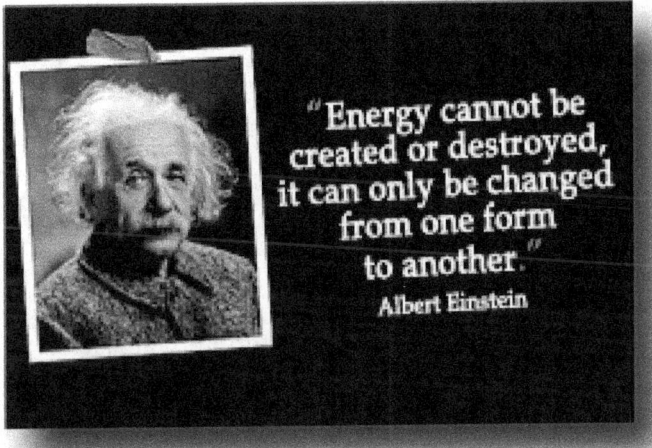

Fig 6: Energy – Can there be Infinite Energy?

Of course, since energy can neither be created nor destroyed, then the energy will need to be transformed into a usable form for the purpose of the activity. For example, solar energy can be converted into heat energy for heating water and other things, while it can also be converted into electricity through the use of photovoltaics. Heat energy can be used as steam to power turbines, which can be used to create electricity to power houses as well as factories and the industry. Mechanical energy can be created through combustion process and the mechanical energy can be used for the forward motion of vehicles such as cars.

Everything on our world both natural as well as manmade runs on energy. If every energy source was to die off on Earth, then life would cease to exist on Earth in a very short period of time. Though through the use of steam power as well as nuclear power, mankind is able to produce electricity to power up all machines and technology: still the main source of energy that Earth and mankind depends on is the solar energy.

We would have to first look at the ecosystem on Earth. The ecosystem depends purely on the sun as all the plants and animals which are part of the ecosystem depend on the input of the sun. Naturally, plants are the first and foremost important part of the ecosystem as they comprise the largest part and they are also at the bottom of the food chain. Every other being that exists on Earth depends on plants as animals that eat plants (herbivorous animals) would be eaten by animals that eat other animals (carnivorous animals). Thus, from a food chain point of view, the herbivorous animals depend on plants to survive and if plants were not there on Earth, then the food chain would break down very fast and an extinction level famine would reign on Earth, Naturally, plants survival depends purely on the presence of solar energy as without solar power, plants would die of very soon and the resulting catastrophe would be global. However, more importantly, due to the photosynthesis process, the creation of oxygen which is critical to the existence of not only humans, but all animals would be hindered if the sun was to stop shining all of a sudden. Hence, energy and specifically solar energy is crucial to survival of life on Earth.

Thus, as it can be understood, nothing runs on Earth (natural as well as manmade) without energy. If there was not sufficient energy reaching to Earth from the sun, manmade energy sources would not work either as water and the environment can freeze up making the production of electricity impossible, since it depends on the combustion process and existence of water. If every water on the planet was frozen with extremely low temperatures, energy production would be impossible (even nuclear energy plants would get effected) and possibly existing technology that we have on Earth would not be able to function at such low temperatures causing the civilization as we know it to stagnate and perhaps die off in the long run.

Of course, with the presence of 21st century technology, we can't really do anything about an extinction event, but we can prepare ourselves to energy crisis which may be manmade. Naturally, the existing energy sources are finite and the dependence on fossil fuel has an important effect on the planet as almost all transportation services depend on fossil fuel. Although there is lot of research going on related to creating electric cars for ground transportation and looking at other options for aviation, it can easily be said that these are still very far away from becoming a reality in terms of feasibility. The reason for may wars has been oil and it seems that reason is still continuing as it can be seen in the world affairs. Thus, it can be said that the lack of energy resources in the world and the competition for these finite sources of energy seems to be a great factor in effecting the fate of humanity.

Of course, on the other side, imagine that there are endless resources available for humanity. Naturally, this will have an effect as to creating so much energy, so that people would be able to do whatever that they want since the whole world economy rests on the competition for resources. In a way, it would make for a world where there are no wars and no political conflicts, but at the same time it would create stagnation in mankind, since without competition then the motivation to succeed is also effected and this would have the effect of creating a static society. Many experts would agree that this would be counterproductive and that a heathy combination of motivation and conflict has to be mixed with prosperity to reach an ideal situation.

Luckily, there are some new sources of energy that are being brought up that can have an effect of resolving these issues. Fusion energy certainly seems to be a good way to create unlimited sources of energy, as the energy yield from a fusion reaction is very high. Technically, fusion is the same way that our sun creates energy, so you can surmise the amount of energy that can be created by using fusion. However, the biggest issue with that would be the fact that creating such large amounts of energy is difficult and containing such a heat would be a humongous task to complete as you would need to deal with thousands of degrees of heat which would be troublesome as best.

Naturally, fusion itself also leads to radioactivity which also needs to be contained as well. There are also other sources of exotic energy such as antimatter, which is potentially an endless source of great energy as per the famous novel "Angels and Demons". When antimatter comes in contact with real matter, they annihilate each other and great amount of energy comes out in proportion to Einstein's famous equation $E=mc^2$. Though we still seem to be very far from having the technical ability to master antimatter, it can be a serious gamechanger that can change the fate of the humanity but also the fate of a single human being: **YOU**.

CHAPTER VI: COSMOS – CAN HEAVENS PREDICT OUR FUTURE?

As stated in the beginning of this book, cosmos has been a very powerful dynamo for the mankind. Whether you look at it as from a point of view of astronomy or from astrology doesn't matter as both can come to the same conclusion that they can affect your future in some way. Well then let's take a look as to how this can happen. Is it really possible the changes in the cosmos effect our future? Are horoscopes for real? Can the heavens really determine your fate? These are all very good questions and for eons people have believed in these that their fate lies in the heavens and in cosmos.

There are three different theories that can apply to this situation. One theory would be to say that everything happens randomly in the universe and thus there is no situation where the sky or planets or stars would have any bearing on your life. Everything happens in the universe independent of each other and the law of randomness guides most of the events in our lives. In this situation, there is no need to try to make any sense of any correlations and the thought that our fate can be understood by looking at the stars would be absurd.

Fig 7: Cosmos – Does it Control Our Fate?

A second theory that may be applicable to the situation would be that there is a causality in the universe and every action that you take leads to a certain reaction and your reaction to each reaction leads to another course of action. As complex as this may sound, basically it says that everything in universe is connected to each other through a causality network and thus each result (or reaction) can be predetermined by going back and creating a map of causality (in other words mapping each action and its subsequent reaction to create a map of present and a predictive map of the future). This is similar to what Isaac Newton tried to do as he believed that if you knew the position of every particle in the history of time, then you could create a perfect picture of the state of the universe and based upon that information, you would know everything there is to know about the past and the present as well as the future.

However, this theory goes further and also incorporated something called the Butterfly Effect which is something that comes from the Chaos Theory. In this theory, Chaos predominates over the universe and a simple act can have dire consequences on an unrelated act. As a classic example, it is said that a butterfly flapping its wings in Japan may cause a hurricane to happen in Chile in the Southern Hemisphere. While classical physics may say that this is impossible, chaos theory states that even the smallest reaction can have unintended consequences and hence the presence of chaos determines the state of any system.

Contrary to popular belief, there are ways to predict the behavior of chaotic systems in mathematics and if you combine this chaotic systems theory with the deterministic theory, then you can have a pretty good idea as to how to predict the future of any event. While this may sound to be a very complicated event, actually it is easier than it sounds and if you had a large enough computer (say a quantum computer), then you could create predictive models as to how you can determine the future of any system (including your future). Although this may sound like science fiction, actually this is used from predicting the weather to even predicting potential earthquakes all around the globe and can be a gamechanger.

So then how do you incorporate horoscope into all this. How does the retrograde of Mercury effect electronical stuff? How does having Venus in your sign introduce a new love affair into your life? Why does your career get a boost when you have Jupiter in your sign? Why does the retrograde of Saturn has dire consequences on your financial life? What is the best day to have a job interview? All of these are very good questions and they have been debated over the centuries. Interestingly if you do a survey of a large group of random people across the world, you would see that majority of people believe in horoscopes and as a result they will concur in some common solutions (such as don't buy electronics in middle of Mercury Retrograde). Since the scope of this book is not to focus on the rules of horoscope but to examine the role of heavens in the fate of humanity, we won't go into detail of what each position of the planet means for your horoscope.

However, it would not take such a large lap of fate to say that if we believe in the causality theory as well as the chaos theory, then it would make some sense to say the position of the planet Jupiter or the planet Mercury may have a certain impact on a person or on the whole planet. Without knowing all the variables, it would be difficult to discern each effect, but based on a large pool of people's experiences, it would be safe to say that some commonalities can be accepted.

For example, let's take the planet Jupiter. It is astronomically the largest planet in the solar system and thus has a large effect on the whole solar system as it exerts a certain pull on all the bodies in the solar system. Its presence disrupts the whole solar system and every particle or object in the solar system would be affected by Jupiter one way or the other due to Newton's theory of gravitation. Hence, it would be wise to say that the presence of Jupiter in the solar system has a net effect on Earth as well as on any object that is found on Earth including humans.

Though it would not be so easy to create a correlation between Jupiter and your career, it is still possible to say through the Chaos Theory that some effect may be present between the presence of Jupiter and some effect on a person or a group of persons on Earth. It would be impossible now to create a conversion formula from the effect of Jupiter to effect on a person as we would not know all the variables. However, we can reverse engineer such a formula by looking at the effect of Jupiter on a large group of people (i.e those who constantly live by rules of horoscope) and based upon their feedback (i.e such as feedback on twitter and blogs about the preciseness of horoscopes), it would be possible to create a collaborative theory that Jupiter has a certain effect on finances and career.

Hence, it can be said that there is some correlation between the positions of objects in the universe and the fate of humanity (though not necessarily just by horoscopes but from actual theories like the Chaos Theory). If we have a large data pool, it would be possible to predict the outcome of each event on each person or on group of people on the planet. Of course, this prediction would also have to take into account the various ideas that have been presented on the above chapters. Entropy, energy, chaos theory, Gambler's Fallacy, the arrow of time and the position of objects in heavens can come together to create a very large ecosystem where the fate of the humanity and its each subset (ie you) would be effected by these systems and understanding them fully can help predict the future of societies and its members such as you.

CHAPTER VII: HOW TO PREDICT OUR FUTURE

We have discussed several important things that effect the future of humanity as well as its future fate. These can be summarized as:

- Arrow of Time
- Entropy
- Energy
- Gravity
- Chance
- Cosmos

As it is seen in above chapters, all of the above criteria effect the future of humanity and they have shaped civilization to the way as we know it. As each person is also a subset of civilization then as a result, each person also gets effected by these powers to be. So, then you may ask yourself as to how you can shape your future.

Fig 8: Cosmos – Can We Predict Our Future?

First of all, you need to understand that you live in a world that is dominated by entropy, dwindling energy resources, time arrow that only moves forward, limitations in gravity that makes Earth your cradle, law of chances that doesn't necessarily support you as well as cosmos which may have an undue influence on you.

While you may be instinctively aware of these constraints above, it is important for you to shape your future accordingly. You need to always make your decisions in such a way to increase your chances and you need to stay away from making decisions as per Gambler's Fallacy. You will have to think of the worst, but you will need to plan for the best. As you know from above chapters, there is no perpetual motion machine that exists and thus you will need to realize that this principle extends to your life as well. In your life, you will also have to keep adding energy and work to your life so that whatever outcome you are hoping for can continue to generate results for you (i.e. Working for salary so that you have some money to spend for your daily needs such as food or transportation etc.)

You will have to take into account entropy so that you can know that everything in life will eventually fall apart as the universe aims you toward chaos rather than an orderly life. Hence, you will need to be aware of this basic principle and you will always have to take precautionary measures to overcome the destructive effects of entropy in your life. Some examples of this would be to live healthily with exercise and dieting to reduce the effects of entropy in your body, to put some of your money into different modes of savings so that you have money stored away for bad days, to have a backup job available in case you lose your primary job and to have more than one skill to help you survive under different living conditions. With proper planning for your future taking into account the 6 criteria detailed above, you may be able to overcome most difficulties in life and you will be able to mitigate any negative effects so that you can be the master of your own destiny. While this may require some trial and error, you will be abşe to triumph in the end.

CHAPTER VIII: WHAT DOES FUTURE HOLD FOR US AS HUMANITY?

Civilization or humanity is formed as a unification of all people on the planet. There are nations where certain group of people (tribes in ancient civilizations) regroup themselves, but this doesn't mean that we don't have many things in common. One thing that can be said for everyone is that everyone is born in 9 months and 10 days and all human beings cry when unhappy and all laugh and react in joy when they are happy. Whether you are Anglo-Saxon or black or yellow or African or Asian or from Amazon doesn't matter as all humans react in general in a same manner to similar stimuli. This itself shows that in a way humanity is on the same boat.

Fig 9: Travel to Stars and Beyond: Is it Possible?

If you look at all the science fiction movies that have been made since the beginning of science fiction movies, you will always see some common patterns evolving in them. One such common pattern would be the ability of humankind to expand into stars. Since the time of Jules Verne, mankind has dreamed of going out of Earth to other destinations. Since Moon is nearest, it has been the first destination that comes to mind for out of Earth destinations. Then Mars and Venus have been the near second choices for destinations beyond the Moon. Many futuristic science fiction books will show that mankind is going to stars beyond our solar system such as Alpha

Centauri and even beyond to far distant stars such as Barnard's Star, Tau Ceti et. There are many stories written by science fiction giants such as Isaac Asimov and Greg Bear and many others where humanity has spread around the stars and has formed a vast empire.

In fact, even the most famous and most popular science fiction TV series of all times, Star Trek talks about the United Federation of Planets, where Earth is an important member and as center of the Federation it commands a galaxy level coalition of planets with various alien races such as the Vulcans and the Andorians to support the Federation. Of course, while this may be a utopic situation, humans as a whole has the drive and the inclination to go where no man has gone before and one day such a federation may become a reality. While Star Trek happens in the 23rd century, many scientists claim that traveling and colonizing the stars may become a normal occurrence in couple of centuries.

If we look at the world history, we see a similar pattern that happened in the past so that humanity spread farther away from Europe to other continents like Australia and America. Today perhaps the United Nations can be thought as a very simplified version of the United Federation of Planets in Star Trek. The most basic drive of humanity is to expand outward and to create new resources. This is the main reason why humanity was able to expand so aggressively in the world in a relatively short period of time. Humanity has always looked outward to expand and to have more land. It is safe to say that perhaps in a decade or two, the moon will be certainly conquered, and Mars will be the next destination. Furthermore, by end of 21st century, it won't be wrong to assume that Jupiter and the far reaches of the Solar system will be inhabited or colonized looking at the reflections of the past.

Once the solar system is colonized, then it can be theorized that in 22nd century mankind will be going to further distances and possibly Alpha Centauri, which is the closest star system will probably be colonized and it will pave the way for destinations to far reaches of our galaxy by the 23rd century. This is the basic premise of humanity and inevitably it will happen one way or the other.

Of course, while long distances are being colonized, another important caveat that humankind has to master for its future would be energy. Fossil Fuels and even renewable energy resources are not enough to put humanity to the next level that is needed for a true technological revolution. In order for this to happen a clean source of energy needs to be utilized. Nuclear Energy from fission has possibilities but it is not clean enough and energetic enough for extreme amounts of energies that will be needed.

One such good possibility would be to use Fusion Energy. Fusion has been discussed as a potential source of energy for many decades, but problems in containment of the heat was a major issue as the heat reaches thousands of degrees in a fusion reaction, similar to the sun which also produces heat and energy through fusion. With a proper fusion reactor, it will be possible to produce endless amount of energy that can fuel up the next cycle of civilization which is needed to take humanity forward to the next step. Naturally, the heat containment problem can also be controlled with advancing technology in materials and especially in nanomaterials. Furthermore, the use of magnetic fields will also be useful in creating a situation where the heat and energy can be contained properly with safety.

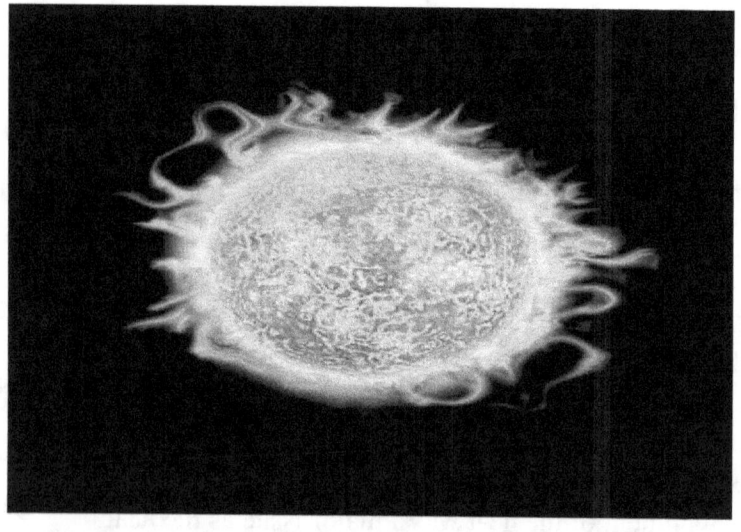

Fig 10: Fusion: Is it Answer to Earth's Energy Needs?

There is a special type of Fusion called Helium 3 Fusion where a special Helium 3 molecule is used as a fusion fuel. The importance of such a Fusion system is that it is called neutronless fusion where you would not be faced with the radiation byproducts of a classic fusion reaction. Furthermore, the containment of such a reaction would be more easier as compared to classical fusion.

Many nuclear energy experts agree that having a Neutronless Helium 3 fusion would be a solution to Earth's energy problems as we can generate enough electricity to handle all of the needs of humanity. Electricity is the basic cornerstone of all technology and thus everything from production to transportation can be handled with electricity and both domestic as well as industrial uses would be beneficial for the future of mankind. The only problem would be the fact that Helium 3 fusion can't be done on earth due to limitations in finding Helium 3, but the good news is that large amounts of Helium 3 exist on the moon which makes it interesting from futuristic point of view. Moon is already within the reach of humanity and thus mining for Helium 3 will be a great incentive for mankind as two birds with one stone can be hit. By going to the moon and having a lunar outpost there, mankind can finally have his stepping stone to the Solar System and to the stars and by mining Helium 3, it will be possible for mankind to have limitless amount of energy to use on Earth as well as for space exploration.

Hence, Helium 3 and the Moon can be important milestones in the future of humanity so that that both the energy needs and the expansion need of humanity can be met in a safe and reliable manner. Of course, more exotic means of energy production such as using antimatter will be next as that has even more of an energy potential then Helium 3 Fusion. In addition, designing a propulsion system using an antimatter drive can be even more beneficial for mankind to escape from Earth and to colonize the Solar System and even the stars. Having limitless energy will also have the added benefit of helping humanity combat entropy as well since disorder can be best combatted with energy which can be used to create order.

Of course, as humanity progresses, it may even be possible to explore more advanced forms of energy such as zero point energy extraction. Theoretically, extraordinary amount of energy exists in each particle in the universe and this energy is called zero state energy. Using this energy itself is called zero-point energy extraction and it can be useful in having an unlimited amount of energy from any point in the universe. You can imagine that with such limitless amount of energy, it will be easy for mankind to achieve anything and even the concept of money can disappear overnight if such limitless amount of energy can be tapped. If we think of how advanced 3D printers have become, theoretically with limitless energy you can replicate any matter and have anything you want free of charge. In one way, humans will become Gods overnight if this energy can be tapped.

Once humankind can master energy then the Basic Law of Thermodynamics which says that Energy can neither be created or destroyed will become obsolete as humankind can have energy at will to use as he pleases. In combination with advanced technology, this will pave the way for humanity to explore the far reaches of our galaxy and perhaps even to other galaxies such as Andromeda and this can cause a significant advancement for human civilization.

Of course, the final criteria that needs to be discussed is the ability of humankind to store and process information. As many historians' state, civilization started to leap forward with the advent of writing which for all intents and purposes started with cuneiform which was a preform of writing. This way historical records could be kept and as a result, this allowed a legal system of rules and regulations to be formed so that the civilization could go forward. Having rules and regulations that are written and passed from generation to generation can be very useful since it will allow some semblance of order which can be used to combat entropy or disorder in society. Moreover, with writing, historical records as well as literature was preserved so that the essence of human society could be passed to future generations.

With the advent of Alan Turing, the concept of computing and processing information started a new age in the dawn of humanity. Using computers, the concept of digital recordkeeping and digital information processing entered our lives. While this was mainly a government effort in the beginning, eventually it turned to a more simplified process that can be easily used by a simple man. Eventually, the concept became more common fold and anyone could use Microsoft Word to jot down his thoughts or use Excel to calculate his finances or use Powerpoint to make a presentation to explain their ideas. Now more records and information exist at this time then all of the past history of humanity combined. Now our future generations have access to unlimited amount of data and records to reconstruct the past.

Fig 11: Quantum Computer: Unlimited Data Storage and Processing

As technology advances and as energy problems become obsolete, this will cause more and more advanced quantum computers to be built, which will pave the way for instantaneous record keeping and data processing. Such advanced data processing capabilities will open the way for instant analysis of any situation which will in turn open the way for understanding any problem and creating an instant solution for it. It may be possible that in 2 centuries, we may have such advanced quantum computers that we may be able to store information in the fabric of space time itself. This would be a great way to overcome many limitations and along with zero point energy extraction, this would allow humanity to manipulate space time continuum itself.

Thus, it can be summarized that humanity is face to face with wonderful developments and in 2 centuries or less, it can be theorized that humanity will reach heights never thought of before and humankind will manipulate space and time continuum. As it can be seen in past history, once humanity starts going forward then many things will start happening at once as some events eventually lead to other events to unfold. This is seen in Renaissance as well as in beginning of 20th century as we had extreme developments such as the first flight by Wright Brothers, first car by Henry Ford, first lightbulb by Thomas Edison, first wireless communication by Marconi, first telephone by Alexander Graham Bell, first AC technology by Tesla etc. These were all wonderous developments that happened in a relatively short period of time of few decades and these inventions still define the world that we live in today. The author of this book believes that a similar period of perhaps 30 to 50 years is about to start where inventions will start to happen one after the another, so that wonderous developments related to unlimited energy, transportation and quantum computing will come about. As suggested above, these developments will lead to even more advancements in a short period of time.

Hence, a wonderful time awaits the future of humanity regardless of all the bad news we seem to be seeing when we watch or read the news. This is partially due to the fact that humankind is geared towards getting up regardless of the calamities that it has faced. You can see that even after two devastating world wars, humanity was still able to rise up and create the 21st century civilization that we enjoy today. Perhaps you as an individual may not be able to see the results of these magnificent changes, but it will still be magnificent and will be enjoyed by your grand-grandkids.

So, we must close the book by saying **Godspeed for the future of the humanity**. Perhaps the outlook may look gloomy with the state of world affairs but all the factors that has hindered the future of humanity from progressing will eventually swing the other way and they will accelerate the advancement of humanity to a new era of Wonders. So buckle up and get ready for the ride.

www.ingramcontent.com/pod-product-compliance
Lightning Source LLC
Chambersburg PA
CBHW070458220526
45466CB00004B/1877